Utterly Remarkable Facts About The

Human Body

by
Ryan C. Burke

TABLE OF CONTENTS

INTRODUCTION

The human body is a splendid work of art, but it's not actually a single entity. Instead, this marvelous machine is an organized network of approximately 50 to 100 trillion cells. These cells are organized into specific structures and have evolved to be best suited for a very specific purpose. Cells in our special senses, such as our eyes and ears, could perhaps be analogous to language professors, such that they have the ability to respond to information in the external environment, and translate that information into electrical information, which is how the brain communicates. Cells in our lungs could be analogous to stock brokers, who constantly maintain the exchange of oxygen and carbon dioxide, along with other gases. Regardless of the role, somehow these trillions of cells work together for the greater good of the whole, which we label, me.

The following is a preposterous look at some of the remarkable feats and properties of the human body. First you must be WARNED! You are about to go where no one has gone before! At least, not anyone who had any kind of social life. I mean we're talking saliva-filled canyons, dinosaur flatulence, super-nuclear diets, psychokinesis, and much more! So if you're among the brave who are ready to take a long, cold look at the real world...well the really weird one anyway, then consider yourself warned.

WHAT IF YOU STORED ALL SALIVA EVER SALIVATED?

Close your eyes and imagine yourself inside a small bakery. The warm smells of rising breads and baking sweets brush softly against your nose, and the smell triggers immediate salivation. Our autonomic nervous system is responsible for stimulating the various salivary glands, causing them to secrete their enzyme-rich materials onto the surface of the tongue.

This liquid, which can be extremely gross if it's attached to your ninety pound Labrador, serves a remarkable purpose. Although more than 99% water, saliva also contains small amounts of mucus, ions, and a few enzymes involved in digestion and the immune response. This lubricating fluid also helps moisten the esophagus when swallowing.

I'm not sure if anyone knows how much wood a woodchuck can chuck, but on average, we do know that each day you will secrete anywhere from 1000mL-1500mL of saliva[1]. That's the equivalent of 2-3 water bottles of saliva per day. Oh yeah, and where is it all going? We swallow almost all of it. That volume also equates to 4-6 cups, and since it's pretty much all water, does that count toward our daily requirements?

Let's take this one step further. If the average human lives around 75 years, how much saliva do you suppose that would be? Well, if we take an average, we secrete 1.25L of saliva daily. Yummy. Multiplying that value by 365 days per year, and again by 75 years per lifetime, we arrive at an astonishing value of, 34, 218.75 Liters of saliva (an extra 25 liters if you account for leap years!).

If you were, for whatever strange reason, curious about the amount of human saliva that has ever been produced, well you're a weirdo. But that's okay because so am I! Current estimates by the Population Reference

Bureau place the number of humans ever to live at around 108 billion[2]. An astonishing number, but even more astonishing is the amount of saliva that has ever been produced, and that amount is on the order of 3,695,625,000,000,000 Liters (3.7 Quadrillion liters).

One liter of water is equal to 1/1000 (0.001) cubic meters. If we try and visualize this disgusting volume of slobber, it would work out to fit nicely in a cube measuring 2.2 Trillion meters * 2.2 Trillion meters * 2.2 Trillion meters. According to the National Parks Service website, the Grand Canyon has an approximate volume of 4.2 Trillion cubic meters[3]. So, if we could have somehow stored the entire history of human saliva in the Grand Canyon, we could have filled it almost to the top. That would be one hell of a swimming pool!

WHAT IS THE MAXIMUM, TOTALLY NOT POSSIBLE HEIGHT A HUMAN COULD GROW?

Gravity is not just the force that keeps us from floating off into space, it's also critical for determining the shape of matter, and a key ingredient to the formation of solar systems. Thanks to it along with our connective tissue, we don't have to worry too much, but imagine for a moment that all of our approximately 50-100 Trillion cells were lined up neatly, in single-file. How tall could we possibly grow?

The individual cells in our body vary greatly in size, with an average diameter of around 50 micrometers, or 0.00005 meters[4]. To put this value into perspective, an regular sheet of paper is around 0.1mm or .0001 meters. So, if we take our average cell, we can stack two of them to get the thickness of a sheet of paper.

Let's take the middle value for our estimated range of cells in the body, 75 Trillion. At 50 micrometers in length per cell, multiplied by 75 Trillion cells, we have a theoretical maximal possible height (albeit not terribly functional unless you have really high cupboards) of 3,750,000 km. At this height you would extend ten times further than the moon, which would now make a great soccer ball. Heck, who's to say the cow didn't jump over the moon?

WHAT IS THE HALF-LIFE OF

YOUR VIRTUAL NON-REALITY?

Whether you're seated or standing (please put this away for now if you're driving), you are probably thinking to yourself how awesome it would be to swim in the Grand Canyon, if it wasn't spit, or play soccer with the moon. Yet while you thoroughly enjoy reading this book, something quite devious is at work, you're being robbed! Now before you do something hasty and call 911, I must inform you that you are being robbed by your eyes. So, I wouldn't call if I were you because that would suck to have your eyes arrested and taken down to the precinct.

If it isn't already a terrible joke that, in order to function properly, we require 6-8 hours of sleep per night. I mean, if you're a lover of sleep like myself (I should be awarded an honorary P-H-D in R-E-M), you're missing out on 1/3 of your life. But wait! There's more.

On average you will blink between 15-20 times per minute, and this serves to keep the eyes lubricated. During this time a process known as blink suppression occurs, and it's essentially a small gap of 200-250ms where we lose visual awareness[5]. It's a very small amount of time, and our brain has mechanisms to compensate and provide a smooth image.

Moreover, our eyes must constantly move in order for us to perceive a sharp image. These small movements are call saccades, and they are the brain's way of continuously comparing the external environment. During this process, two points are measured, processed and an image is perceived. Another form of suppression, called saccadic suppression, essentially deletes the blurred image between the two points. This type of visual suppression lasts anywhere from 100-150ms, and occurs around 50 times per minutes. To demonstrate this, head to the nearest mirror. Look closely at one eye, then quickly look at the next eye. You can try as many times as you like, but you will only ever get to see one eye or the other, but never the moment in between.

If we take the amount of time lost in a minute due to these visual suppression mechanisms it would add up to approximately 10 seconds per minute. So, not only are we losing 33% of our life to sleep, we are also having an extra 17% of our waking life completely deleted! The terribly depressing truth my friends, is that we can never truly live a full life. At very most, we could only hope to live half a life. May your half life be filled with blessings. I'm going for a nap.

Reduce, Reuse, Recycle.

What Was Once My Fresh

Air Is Now Yours

The Earth's atmosphere is a mixture of gases, most of which is nitrogen, oxygen, and argon, that are bound to Earth due to gravitational forces acting upon them. The approximate mass of our atmosphere, as per NASA.gov is 5.1 x10^{18}Kg[6]. (5 Quintillion kilograms!) At a certain point during the evolution of our planet, the composition of the gases in air became sufficient to support life. It is from this gassy soup that we breathe life.

How much of this life-sustaining nectar do we breathe though? Well, if you consider that we breathe between 10-15 times per minute, and each breath moves approximately 500 milliliters of air into your respiratory system, on average we take in about 6 liters of air per minute[7]. Earlier we suggested an average life to be 75 years, which means that we'll live roughly thirty-nine million (39,420,000) minutes. Not so shabby! Over the course of our life then, we will each inhale and exhale roughly 236 million (236,520,000) liters of air. The conversion for mass is 1.286 grams per liter, at standard atmosphere and temperature, which means we breathe around three hundred thousand kilograms (304,164,720 grams) of air in our life.

Many practitioners of meditation reach heightened levels of blissful awareness by slowing the breath down. I think they're really feeling blissful because they are less likely to breathe in recycled air molecules! If these gases are trapped to the Earth, as we are, by gravity, and there have been around 108 billion people ever, that works out to be 33 Quadrillion kilograms (3.3×10^{16} Kg) of human-breathed air. Now this is just human breath, but there are even more animals. If we were to take all of the air-breathing creatures ever to exist on Earth, it is quite easy to see that this value will far-exceed the atmospheric capacity. With that stated, it is quite possible that the air you are breathing right now contains trace gases that were farted out by a large brontosaurus hundreds of millions of years ago. If that doesn't give a new meaning to "*a breath of fresh air*", I don't know what would.

Barbecue TNT & Me: The World's New Super-Nuclear Diet

The average adult requires approximately 1200-1800 calories per day of nutrition[1]. We often look at calories as these evil villains that try to invade and multiply in your abdomen, but the calorie is simply a unit of energy, where 1 calorie equals 4.184 joules. It's not actually energy that's the antagonist here, rather where you derive your energy from, since we are merely a conglomeration of cells, and these cells have required protein, carbohydrate, fat and vitamin values in order to function properly.

The human body is quite a marvel. Have you ever wondered how much energy it takes to fuel your body over its lifetime? If we use the average of the above range, 1500 calories, that would be equivalent of 6, 276 joules of energy per day. In a day there are 86, 400 seconds, which means that we require .073 joules per second. A joule per second is by definition a watt, so the average adult runs off 73 milliwatts. That means we are 1370 times more efficient than a 100 watt light bulb, or if you're a glass is half empty type, 1370 times lazier than a 100 watt light bulb!

So, it doesn't take a whole lot of energy to operate a human being for a second, but we live quite a few seconds. In fact, we live on average greater than two billion seconds. So, if you live until 75 (hopefully you live much longer), you will have consumed close to 180 million joules of energy. That's quite a bit of energy. If now we take into account the entire Earth's one hundred and eight billion humans ever to exist, we will consume roughly 1.94×10^{19} joules (19.4 Quintillion joules).

Not to diverge, but nuclear bombs have a remarkable ability to generate tremendous amounts of energy in small periods of time. Imagine for a moment there existed a capacitor capable of storing the 19.4 Quintillion joules of energy, and one day some crazy person decided to ignite it. What kind of explosion would you expect?

The single largest explosion from a nuclear weapon occurred in 1961, when the Soviet Union dropped a 50 Megaton bomb, called Tsar Bomba, over northern Russia. According to tsarbomba.org., the blast was so massive it's mushroom cloud extended 65 km into the sky, and had a diameter of 40 km$_8$. If we consider an average human could fit comfortably in about two and a half square meters, half of the Earth's population could easily fit inside the mushroom cloud. Of course, the roughly 4000 degree Celsius temperatures inside may not be for everyone. The amount of energy released from that explosion was 0.21 Quintillion joules, meaning humans have consumed enough energy to cause an explosion with 92 times the energetic release. To put this into a slightly clearer picture, that is greater than the combined explosive energy of 128,800 of the bombs that ravaged Hiroshima and Nagasaki.

So good people, please do your part to prevent nuclear destruction! Remember to exercise and poop frequently to ensure that you are giving as much energy back as you are taking in.

Dancing Photons &

Martian Optical Illusions

Remember earlier when I told you that your brain and your eyes work together to delete 17% of your day without you ever knowing? I hate to be the bearer of bad news, but there's more. They are also causing you to hallucinate. Can you believe it...the nerve! (Please make note of the splendid pun...the nerve...brain...haha...look, I'm not very funny, so on the off chance I hit one out of the park, I have to make sure everyone knows). Humans are estimated to have the capacity to perceive around ten million colors. Incredible indeed, but what is color?

The photon is the functional unit of the electromagnetic spectrum, meaning that all radiation on this spectrum occurs when photons vibrate at a specific wavelength. The visible spectrum lies in a small region between ultraviolet and infrared. The associated wavelengths have a range from approximately 400-700 nanometers (one nanometer is one billionth of a meter), which means the cells in our retina are only capable of responding to photons with wavelengths in that range[9].

Let's take the extreme colors in our visible spectrum, blue and red. They have associated wavelengths of 400nm and 700nm, respectively. Since the wavelengths are different, the speed at which they travel must be as well. Now suppose you have a Martian friend holding two super-powerful flashlights, one blue and one red. You guys set up a time for a really neat experiment where he points them directly at you and turns them on at the exact same moment. As reported by nasa.gov., the average distance from Earth to Mars is 228 billion meters[6]. That means blue light would require 5.71 Quadrillion (5,710,000,000,000,000) versus red's 3.26 Quadrillion cycles before they would reach you on Earth. The amount of time each cycle would take can be calculated from the inverse of the frequency, and results 1.3 versus 2.3 femtoseconds (0.000 000 000 000 001 300).

Using these values, we can finally determine how long it would take for the photons travelling from each flashlight to reach your retina. For blue light, it would take 742.3 seconds, whereas the red light would take 750.8 seconds or more than eight seconds later.

So there you have it, once again you have been deceived. What you thought was color turns out to be mischievous dancing photons, vibrating at different frequencies to deceive your retina. It may be helpful, but it's still a deception! Now if people would have thought of this near-useless fact long ago, they would have made stop signs blue so people would see them sooner.

Your Honor, I'm disputing this ticket! Maybe if the city wasn't so obtuse and had blue stop signs I would have had time to react. I mean, do the math! - never spoken by anyone anywhere

IF THE PEN IS MIGHTIER THAN THE SWORD, IS THOUGHT MIGHTIER THAN THE PEN?

I'm sure, on at least one occasion, we've all stared sternly for extended periods of time at a pen or pencil, fully intending to move it with our mind. If not, me neither... I'm just saying...But it would be cool though, right? Unfortunately, many illusionists, despite being super-cool, have misguided the public into believing that psychokinesis/telekinesis is simply a trick. Making people believe that you have moved something using the power of thought is far less impressive, however, than actually doing it. Could thought really have the ability to influence matter at a distance?

First, let's analyze what is required to move, for example, a pen. The pen rests directly in front of you on your desk. Because of the composition, specific wavelengths are reflected and interact with your retina. The retina takes light energy and converts it into electric energy, which is the currency of the brain, yet it hasn't moved. So the pen has actually affected matter (your retina) at a distance without touch. Next, you get an incredible idea that you have to write down. That idea sends electrical impulses down to your arm into your hand which are converted into mechanical force, allowing you to pick up the pen. Your brain influenced your hand, but it never touched it.

Those may seem like silly examples, but they are no different than psychokinesis. The only real difference is that we have textbooks that describe the mechanisms allowing us to see, or how we move. Since we don't have a known mechanism for moving objects with thought, people must generate their own, thereby entering the realm of belief rather than fact. I'm not one for believing what others think. I'd rather measure it, so let's give it a try.

In order to move an object, a force must act upon it. Thanks to Sir Isaac Newton and apples, we know that force is equal to the mass of an object multiplied by its acceleration. Moreover, if we know the distance we are planning to move the object we can determine how much energy is required in joules. Therefore, if we want to move a 5 gram pen over a distance of 1 meter in 1 second, the energy required would be a mere 5 millijoules. That doesn't seem like much at all compared to our nuclear appetite, but how much energy is involved in thought?

The brain uses charge as a form of currency, such that when a threshold is reached, a transaction occurs. We call this transaction an action potential. Essentially there are tiny ions floating throughout your interstitial fluid as well as inside your cells. Some of these ions have positive charges while others have negative charges, and because they are separated by the cell membrane an electrical gradient is formed. When the threshold is met, specific channels open allowing the neuron to depolarize, and an action potential results. The electrical gradient is measured in volts, and a single action potential produces a change of 120 millivolts[9]. Energy can also be measured if we know the voltage and the electric charge. Since the latter is a constant, $1.6 \times 10^{-19}C$, the energy of a single action potential equals approximately $1.9 \times 10^{-20}J$. A very small amount of energy. Contrary to what many of you have been told, we do not use 10% of our brain. In fact, we use 100% of our brain 100% of the time. Now before you panic and assume that you're now trapped in a dead-end job for the rest of your life because some guy wrote a book saying you have used 100% of your brain and could never hope for more, breathe. The truth is our brain is plastic, meaning it is constantly pruning and creating new synaptic connections between neurons, so we are always capable of new memory formation and task acquisition. What was meant was that all cells require metabolic activity for survival, even if it's to a lesser degree, such as in a hibernating animal. Thus, all cells will always be active, at least while we're alive.

Our glorious brains have approximately 100 billion neurons, and these cells display a range of activity from 0.5 to over 100 Hertz, with an average of around 10 Hertz[9]. This means that the average neuron generates approximately 10 action potentials per second. With a single action potential generating 1.9×10^{-20} joules, this means that we could theoretically generate 19 nanojoules (1.9×10^{-8}J) per second, or 19 nanowatts. Well, that's a bit of a bummer! Considering it would take almost four million of you to move a pen, you should probably not spend too much time straining.

Just to leave you with a bit of hope, consider that pen is actually a small tumor. I'm not a religious man, but this could give mathematical validity to the power of prayer. If enough people concentrate on that tumor, ample energy would exist to possibly affect it.

GROWING PAINS

Have you ever caught yourself almost breathless after gazing into the innocent eyes of a newborn? *What a miracle*, you thought to yourself. It's true, the idea that two people can actually combine genetic information to form a new being is remarkable. Perhaps just as remarkable is the rate at which these little ones grow inside the mother's womb. Within approximately 280 days a beautiful 3.5 kilogram baby is welcomed into the world. That means the baby is actually growing by approximately 12.5 grams per day, or about a tablespoon of sugar each day.

That doesn't sound too bad, but then again I'm male and will never have a living being grow inside me. And I'm at peace with that. Inside the woman's womb, however, things are happening a lot faster than you might think. If we consider an average cell having a mass of approximately 1 microgram (0.000001 grams)[4], then the 3500 gram cute little bundle of cells would have roughly 3.5 billion cells. Not that it's a linear process, but this would imply an average of 12.5 million cells per day, or 145 cells per second. When stated like this, it sounds a bit more impressive than a tablespoon of sugar.

Although quick, these tiny bodies grow even quicker during the first year of their lives. According to the Mayo clinic's webpage, the average baby can grow up to three times their body weight in the first twelve months of life[10]. So, the 3.5 kilogram baby can grow an additional 7 kilograms in their first year, which works out to be 19.2 million cells per day, or 222 per second. Now that's pretty quick for little things that sleep and eat their entire day.

TOUCH AND PAIN FROM A

GIANT'S PERSPECTIVE

You are about to embark on an epic battle. The winner of this incredible race will have the opportunity to meet with the mysterious genie who will grant them power to influence a giant! If you accept this extraordinary challenge, you must choose your carriage; pain or touch? Our central nervous systems are extremely organized, and the information entering the spinal cord ascends separately. It only becomes an organized perception once the cortex becomes activated. Information pertaining to pain and touch are quite different. For nervous tissue dedicated to touch, the axons tend to be thicker in diameter and are also insulated by a fatty sheath called myelin. These properties are contrasted in pain receptors, which tend to be much thinner in diameter and contain little to no myelination. As a result, these nervous tissues actually conduct electrical impulses at different average velocities; approximately 10 m/s for pain, and 100 m/s for touch[9].

So the brain perceives touch ten times quicker than it does pain. I'm sure we can all remember times where we have stubbed our toe, or cut our finger, and perhaps noticed that we felt it slightly before the pain started. If not, it's a real hoot and you should definitely go stub your toe. For example, using the treacherous toe stub, the average individual would feel it in 15-20 milliseconds, but the pain wouldn't be perceived for 150-200 milliseconds. Although a small difference, I did promise an epic battle, so let's head to the moon where we can watch this race as it speeds around something much larger, the Earth.

Our home has a massive circumference of approximately 40 million meters. If a giant could lay all the way around the planet, so his head touched his feet, and we took a huge sledge hammer to his big toe, it would take him eleven hours to become aware of it. Although aware of something, it would take more than four days before he realized how much it actually hurt! So ladies and gentlemen, please don't harm giants. If you prick them, they will bleed. It just may take a while for them to realize it.

REFERENCES

1. Tortora, Gerard J. and Derrickson, Bryan. *Principles of Anatomy and Physiology.* 12th. Hoboken : John Wiley & Sons, Inc., 2009.

2. How many people have ever lived on Earth? *Population Reference Bureau.* [Online] 2011. www.prb.org.

3. Grand Canyon. *National Parks Services.* [Online] 2013. www.nps.gov/grca.

4. Sherwood, Lauralee and Kell, Robert. *Human Physiology: From Cells to Systems.* Toronto : Nelson, 2010.

5. *The effect of blinks and saccadic eye movements on visual reaction times.* Johns, Murray, et al. 2009, Attention, Perception & Psychophysics, pp. 783-788.

6. *NASA.* [Online] 2013. www.nasa.gov.

7. *Tsar Bomba.* [Online] 2013. www.tsarbomba.org.

8. Kandel, Eric R., Schwartz, James H. and Jessel, Thomas M. *Principles of Neuroscience.* New York : McGraw-Hill Health Professions Division, 2000.

9. Chuddler, Eric H. Brain Facts & Figures. *University of Washington.* [Online] www.faculty.washington.edu.

10. Hoecker, Jay. Infant and Toddler Health. *Mayo Clinic.* [Online] 2013. www.mayoclini.com/health/infant-growth.